Technical Report Standards

How to prepare and write effective technical reports

by

Lawrence R. Harvill, Ph.D.

Chairman, Department of Engineering
University of Redlands
Redlands, California

and

Thomas L. Kraft, Ph.D.

President,
KVM Engineering, Inc.
Houston, Texas

FIRST EDITION

Copyright © 1977 by Lawrence R. Harvill and Thomas L. Kraft. All rights reserved. No part of this publication may be reproduced, stored in a retrieval system, or transmitted, in any form or by any means, without prior written permission of the author and publisher. Manufactured in the United States of America.

ISBN 0-930206-01-0
Library of Congress Catalog Card Number: 77-70964

Second printing, 1978

M/A Press
P.O. Box 92
Forest Grove, Oregon 97116

This book is dedicated to Estelle (Dorsey) Ratner, whose inspiration and support several years ago motivated us to initiate our first efforts on this project.

"The Gold Dust Twins"

CONTENTS

1.0 **INTRODUCTION** 1
    1.1 Explanatory Comments 1
    1.2 Outline of a Typical Report 2
        A. Introduction 2
        B. Main Body 2
            1. Experimental set-up and equipment summary 3
            2. Data 3
            3. Data reduction 4
            4. Sample calculations 4
            5. Error analysis 4
        C. Discussion 5
        D. Conclusions 5
        E. Suggestions 6
        F. References 6
    1.3 General Approach to Good Report Writing 7

2.0 **REPORT FORMATS** 9
    2.1 Introduction 9
    2.2 Recommended Formats 9
    2.3 Letter Formats 10
    2.4 Memorandum Format 11
    2.5 Form Reports 12
    2.6 Article Reports 13
    2.7 Formal Reports 14

3.0 **TABLES** 17

4.0	**GRAPHS**		19
	4.1	Introduction	19
	4.2	Planning the Graph	19
	4.3	Types of Relationships	20
		4.3.1 Observed relationships	20
		4.3.2 Empirical relationships	21
		4.3.3 Theoretical relationships	21
	4.4	Variables	22
	4.5	Size	23
	4.6	Titles	23
	4.7	Axis Labels	25
	4.8	Scales and Units	26
	4.9	Symbols	31
	4.10	Information on Graphs	32
	4.11	Questionable Points and Deletions	32
	4.12	Folding Large Graphs	33
5.0	**ILLUSTRATIONS**		35
6.0	**ERRORS**		37
	6.1	Introduction	37
	6.2	Definition of Terms	38
		6.2.1 Errors	38
		6.2.2 Random errors	38
		6.2.3 Systematic errors	39
		6.2.4 Precision	40
		6.2.5 Accuracy	40
		6.2.6 Blunders	40
		6.2.7 Computation errors	40
		6.2.8 Instrument precision	42
		6.2.9 Instrument accuracy	42
		6.2.10 Instrument sensitivity	43
	6.3	Error Evaluation	43
		6.3.1 Algebraic relations	44
		6.3.2 Functional relationships	50

6.4	Graphical Representation of Errors		51
6.5	Significant Figures		52
	6.5.1	Rounding off nonsignificant figures	52
	6.5.2	Addition and subtraction	53
	6.5.3	Multiplication and division	53
	6.5.4	Using logarithms and scientific notation	53
Suggested References			55

PREFACE

From our years of experience in education and industry, we are aware of the need for a guide to preparing clear, concise and effective technical reports. Such a book need not focus on the proper use of language or grammar, since many helpful references are available on these subjects. Instead, it should clearly set forth the basic standards for report preparation: the accepted formats, the correct approach to data presentation, and the proper use of graphs, tables, illustrations and other elements. All of these are essential to achieving an effective technical report.

As a result, we have conceived this unique guidebook. You will find it is not a detailed exposition, but a step-by-step summary of the primary aspects of writing an effective technical report. It provides accepted standards for the clear, concise presentation of experimental or theoretical data in tables, graphs and other visual displays. It examines preferred methods for error analysis and discussion within the report, as well as for presenting recommendations and forming conclusions. These and other aspects are explored in clear, simple language to help the student, technician, engineer or scientist prepare a complete and well-written technical report.

This book had its beginnings in 1959, when we produced the first draft for use in undergraduate engineering laboratories in the Department of Engineering at the University of California at Los Angeles. From our professional experience since that time, we have found this work to be useful on a far wider basis, in education, the military, business and industry, and the sciences. We have therefore revised it for publication in this expanded form.

Though originally conceived to help students in the preparation of laboratory reports, this book is equally helpful in producing any type of technical report that employs empirical or theoretical data. We hope you will find it a valuable guide and reference for preparing your next technical report.

Lawrence R. Harvill, Ph.D.
Thomas L. Kraft, Ph.D.

SECTION ONE

INTRODUCTION

1.1 Explanatory Comments

Effective communication has always been an important tool in every field of endeavor. It is even more vital today with the rapidly increasing rate of generation of new information. This is particularly true in technical fields, where clarity, efficiency, and conciseness in written communications are essential.

The first basic ingredient of any technical report is the use of good grammar. Several excellent references are available on the proper use of grammar for writers. Beyond this, every report must be developed in a logical sequence. Careful attention should be given to the preparation of figures, graphs, tables and error analysis. With respect to these report factors, very little guidance information is available.

The purpose of this guidebook is to provide a clear, concise reference to these important topics. At the outset, it is important to mention one specific grammatical rule for report-writing: avoid the use of the first-person pronoun "I" in the report. The pronouns "we", "us", and "they" are acceptable, but their use should be minimized to maintain an impersonal style in the report.

Let us review the steps in setting up a technical report. First, decide on the most appropriate format. Once this choice has been made, then the content should be developed. In doing this, strive to keep the audience in mind at all times. It is particularly important to consider the audience's level of expertise and familiarity with the subject. Some reports must be addressed to several audiences, e.g. the project director, sales manager, and top management officer. For students preparing laboratory reports, it is an accepted practice to write or direct their reports to other students at the same level, rather than to the instructor.

INTRODUCTION

In developing report content, the presentation should flow in a logical sequence from beginning to end. In many instances, this sequence follows the order in which the experiment was carried out. A brief summary of a typical report layout is presented here.

1.2 Outline of a Typical Report

A. Introduction——This is the first statement of the report. The introduction is usually a single paragraph and is vital in setting the stage for the entire report. It must contain a statement of purpose which briefly outlines why a particular experiment is*being performed. A few clear, concise sentences are normally sufficient to explain this. The following are typical examples:

> "The purpose of this experiment was to determine the amplification factor of a common emitter amplifier by theoretical and experimental means. The two methods of solution will be compared and analyzed for validity."

> "The purpose of this experiment was to present three distinct quantitative methods of analyzing force fields in thin plates under tension. The visual concept of force fields was to be investigated to facilitate the understanding of loaded plates in structures."

The introduction also should state the reasons for an experiment, its goals, and a brief summary of the procedure to be followed. The introduction, then, should tell the reader what is being reported, why it is being studied, and how the investigation was conducted. Finally, a short review of the findings and conclusions should be included.

B. Main Body —— After the introduction comes the main part of the report, the body. This section should fully present the topic under investigation. For most experimental reports, the content of the body should be developed in the same manner as the experiment or study was conducted. Typically, the body would be constructed as follows.

---

*The use of present or past tense is proper in notebook reports where procedures and observations are being described. Past tense is preferred when the report is written some time after completion of the experiment.

TECHNICAL REPORT STANDARDS

1. Experimental set-up and equipment summary. Present a brief but adequate description of the experimental set-up and procedure followed. In many instances, one or more figures will effectively present any detailed circuits or equipment set-ups in concise form.

An essential part of every experimental laboratory report is a list of the equipment used. This identification is important for two reasons. First, if a question arises concerning the validity of the readings of an instrument, the description in the equipment list provides a means for locating the same instrument for testing. Second, if a question arises as to the validity of the experimental results and a rerun of the experiment is necessary, then the same equipment may be used to duplicate the original test set-up.

Therefore, the equipment must be accurately described. This is usually accomplished by listing the title of the instrument, its applicable range(s), and its serial and/or property number, together with a description of any identifying marks or the exact location if no other means of identification exist. The equipment list should also include an indication of the instrument accuracy, which serves as a reference for the error analysis.

The equipment list may be included in the body of the report if it is not too long. Otherwise, it should appear as an appendix. The following are examples of typical entries in an equipment list:

Oscillator, Audio, 20-20,000 hz, UR sn 6N321, $\pm 1\%$.
Tensile test machine, Tinius Olsen Low-Cap, 6 ranges of 0 - 125, 250, 500, 1250, 2500, and 5000 lbs, located in room 104 Hentschke Hall, $\pm 0.1\%$.
Voltmeter, D.C., 0 - 10 volts, UCLA, 123-E-456, $\pm 2\%$.

List only pertinent equipment, e.g., oscilloscopes, meters, stop watches, or extensometers. Items such as beakers or meter sticks generally are not pertinent and should be omitted.

2. Data. The data of an experiment are the measurements of experimental variables and other observations. Usually, they are recorded initially in tables and later are displayed graphically for analysis.

INTRODUCTION

The data must always clearly identify what was measured. The units of measurement, the magnitude of measurement, scale factors, and full scale and least reading of the instrument by which the data were taken should be noted. Nothing is more confusing during data reduction and analysis than pages of unidentified figures.

Remember these important factors when taking experimental data:

a. Before any experimental data are taken, be fully aware of what is taking place in the experiment. Know what parameters are going to be observed, and through what range they are expected to vary. In other words, before the experiment begins, know as well as possible what is going to happen.

b. When recording data, keep in mind that a sufficient number of readings should be taken to verify the problem in question. Insufficient data may obscure the facts or trends that were to be investigated.

c. Above all, remember that the data must be correlated with the procedure in order to attain their full significance.

3. Data reduction. In most cases, raw experimental data must be reworked in some manner to present a clear picture of the relationships between the various experimental variables. Typically, this is accomplished by plotting the data in graphical form. From such presentations, various empirical relations may be derived or comparisons drawn with theoretical predictions.

4. Sample calculations. If any extensive calculations are a part of the investigation, include a brief sample calculation to illustrate the particular equation(s) used and how the raw or reduced data were employed. If many cases involving the same calculation are involved, a sample calculation for only one case need be included.

5. Error analysis. Perhaps the most important single item in any report is the discussion of error sources. Either in the "Discussion" section or in a separate section, present a quantitative discussion of the errors involved. This should include the types and causes of error and possible corrections. Make a quantitative estimate of the error. If the error discussion appears in a separate section, then some estimate of the overall error involved should also be included in the "Discussion" section for convenience. Where possible, make comparisons to accepted or handbook values to provide a base reference. If there appear to be no errors in the experiments, measurable or otherwise, the reason for this must be clearly stated.

C. Discussion——In most reports, the discussion of results will be incorporated with the body of the report only if they are very brief. Otherwise, results should be included in a separate section entitled "Discussion." Major inconsistencies, statements about equipment failures, and explanatory comments also should be included in this section. It is likely that the writer will arrive at some incidental findings in working toward the experimental objective. These findings also may prove to be more important than the anticipated results or the answer sought in the experimentation. Such incidental results can logically be placed in the "Discussion" or in a separate section titled "Additional Findings."

Though the experimenter should seek to avoid qualitative observations, they should not be ruled out. In conducting any experiment, always remain alert for any unusual or unexpected behavior of the equipment or the item under test. Report any observed deviations as quantitatively as possible.

Questions raised in the body of the notes to the experiment should also be answered in this section. If questions are asked in a separate section of the syllabus notes, however, then they should be answered in another separate section entitled "Discussion Questions" or some other similar title. Note that the answering of syllabus questions in no way eliminates the need for a discussion.

Any pertinent theory also may be presented in this section. Personal opinions may be included if it is clear that they are opinions rather than facts.

D. Conclusions——Specific conclusions should be drawn from the experiment. These conclusions should be based only on the results obtained and must answer the objectives and goals stated in the introduction. Any objectives or goals which were not fulfilled must be carefully explained.

If the writer feels that no conclusions can be made, then the reason for this must be stated. Any conclusions made should be brief and to the point. No new material should be introduced in this section of the report.

INTRODUCTION

It is important to be alert to the potential value of "failures" in any situation. Report any failure which relates in any way to the objectives of the experiment. In many cases, a "failure" is as useful as a success in that it eliminates one alternative or another in the evaluative process. A "failure" also may serve to identify or locate weak points in a system.

Another important aspect the experimenter should keep in mind is that results which wholly contradict the original expectations should not be overlooked or ruled out too quickly. The natural tendency is to attempt to rationalize such results as instrument or experimental error. Whenever no readily verifiable reason exists for such a situation, carefully review the sequence of assumptions and postulates which led to the prediction of the anticipated results. A classic contemporary example of such a circumstance occurred in Van Allen's investigation of the earth's radioactive belts. The instruments had been carefully chosen to indicate values above those which had been predicted from theory and low-altitude experiments. The data telemetered back by the first probes indicated that the instruments were "pegged" at the maximum reading. When this happened, the results were interpreted as "failures" or misinterpreted as bad measurements. As later discovered, the error was in the initial assumptions and was in the range of an "order-of-magnitude." Only when instruments were flown which had significantly higher ranges was the error found.

Itemized listings of the important results are excellent means of presenting the conclusions.

E. Suggestions——Clearly state any suggestions for improvements or refinements, pertinent to the experiment under consideration. Open criticism without suggested action is not considered satisfactory. Recommendations should be supported by sound reasons, and should be both appropriate and helpful.

F. References——In any report, a list of references must be included whenever written materials are used for any purpose. Some typical examples are: reference books, textbooks, other reports, or equipment manuals. For the appropriate format by which references should be cited, consult any textbook or professional journal for examples. Most writer's reference books also include sections on proper referencing.

1.3 General Approach to Good Report Writing

The report you are writing concerns an experiment. Every experiment has a series of common elements.

A. It is controlled. This means there are "standards" or "controls" used in the experiment to ensure operability and validity of the methodology.

B. It involves:

   (a) **determining** a law or relation among variables, or

   (b) **verifying** a known law or relation, or

   (c) **determining** an effect of a known law (or no law), or

   (d) **testing** a hypothesis about a set of variables.

C. It is based on one or more hypotheses about the variables involved.

D. It necessitates one or more assumptions about the behavior of the variables involved.

All of these items point to the importance of establishing a "Measurement Plan" to support the experimental objectives. The contents of such a plan should establish the four items just cited. In preparing a plan for measurement, be sure to address the following questions:

   (1) What variables are to be determined in the experiment?

   (2) Can the variables be measured directly?

   (3) Will the variables be measured indirectly, assuming some law is valid?

   (4) Will the variables be computed?

   (5) What equations or "models" will be used to make the determinations?

   (6) What assumptions are made or implicit in the measurements (e.g., linearity, range, dependence, independence)?

In other words, be aware of the assumptions that are part of the experiment; and, be certain you are measuring *only* what you intend to measure — and that you have eliminated unwanted influences on the data collected.

INTRODUCTION

These points are mentioned to illustrate some of the reasons why experiments fail or have little value. In many cases, an instrument is selected on the basis of availability, and data subsequently collected. The method of measurement or relation used in the instrument may be subject to certain environmental variables, such as humidity, temperature, radio frequency interference, pressure, gas, or magnetic fields. The experimenter must recognize that such influences can negate the data. If the exact methods of measurement are known, these effects can be predicted or countered. If the principles used are unknown, then the experimenter should carefully identify the instrument to qualify the results, and should use known standards to demonstrate that peripheral effects from uncontrolled variables are negligible. It is extremely important to think through what really is being measured and how it is being measured.

One additional point should be heeded when planning measurements. The actual act of measurement — the instrument and the experiments — can alter the properties of the system being tested or measured. For example, putting an electrical instrument in a circuit can alter the behavior of the circuit in subtle ways. Similarly, optical measuring instruments can alter the light system. Thus, use caution in planning the measurements.

Finally, the report-writer should seek to be as clear and concise as possible without destroying the logic and continuity of the overall presentation. Always use precise language. "The temperature was 1250°F" is insufficient. The reader needs to know whether the temperature was 1250°F ± 200°F or 1250°F ± 5°F.

The following sections of this guidebook present detailed information on the use of formats, tables, graphs, illustrations, and error analysis in technical reports. Hopefully, these sections will provide sufficient information to enable you to prepare clear, concise, and effective reports of every kind.

SECTION TWO

REPORT FORMATS

2.1 Introduction

The growing complexity of modern technology has created a demand for ever more efficient communication, particularly in written reports. Greater standardization in report-writing has developed to help bring about this needed efficiency.

Selection of an acceptable format should be the first consideration in preparing a report. Standard formats are prescribed for each of the many purposes of the technical report. For example, a report may be intended to present general or detailed information, to make a specific recommendation, to give instructions regarding equipment operation, or to document legally for purposes of patent protection. Often, a report must fulfill a combination of these purposes. In all cases, the author has a conscientious obligation to select the format carefully and to plan the report. Both the intended reader and the content of the report should guide the writer in selecting the appropriate standard format.

2.2 Recommended Formats

Written formats generally are classified by the purpose of the writing rather than by the content. There are two general classifications of formats: the correspondence format and the report format.

In the *correspondence* classification, there are two types of formats: letter formats and memorandum formats.

REPORT FORMATS

2.3 Letter Formats (see **Figures 1 - 3**)

2.3.1 The letter usually is used for communication with persons outside the author's organization or institution.

2.3.2 The letter should be brief and concise.

2.3.3 The letter generally is confined to *one* major topic. Two distinct topics necessitate two letters. Various routing procedures and a separation of interested readers make it essential to separate distinct topics.

Figure 1

Block Letter Format
(The most common format)

If you are a student or individual, place your address in block form directly above the date.

Figure 2

Indented Letter Format

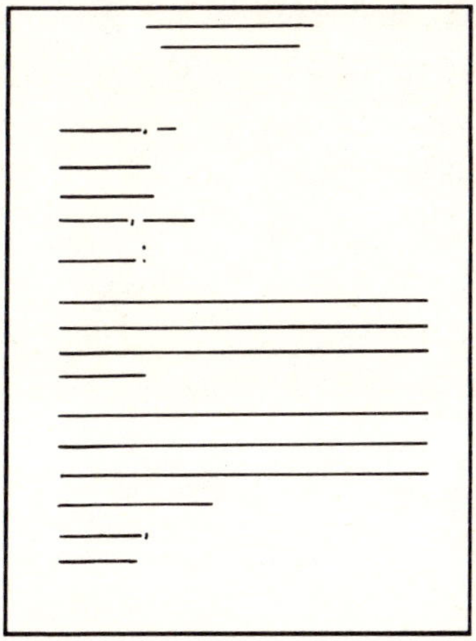

TECHNICAL REPORT STANDARDS

Figure 3

Full Block Format

2.4 Memorandum Format (see Figure 4)

2.4.1   This format is restricted to intra-organization communications. It is recommended for use between only two people, the author and the reader. Outside communications require letter formats.

2.4.2   Brevity and conciseness are the most important aspects of this format.

Figure 4

Memorandum Format

(The inclusion of the word "MEMORANDUM" is optional)

REPORT FORMATS

2.4.3 Each memorandum is restricted to a single topic. Separate memoranda are required for more than one topic.

2.4.4 The entire report should not exceed one or two pages, except in rare instances.

2.4.5 The contents, in order, should be: introduction, statement of problem, discussion, conclusions, and recommendations.

2.4.6 Appendices may be added for graphs, data, or other addenda.

In the *report* classification, there are three types of formats: form, article, and formal.

2.5 Form Reports (see **Figure 5**)

Form reports are generally used in most large organizations. These "forms" serve to standardize the many functions of a large institution. Requisitions and work orders are two typical "forms." Form reports usually fulfill a special purpose and permit little, if any, variation. In this format, brief descriptive phrases are preferred to whole sentences.

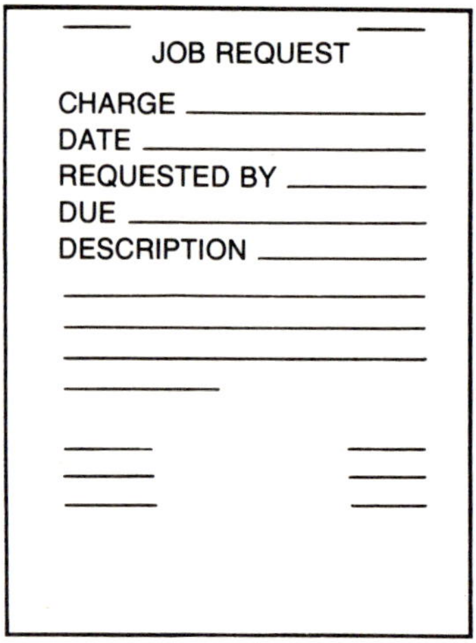

Figure 5

A Typical Form Report

TECHNICAL REPORT STANDARDS

2.6 Article Reports (see **Figure 6**)

2.6.1 This format is normally used in instances where the reading audience is large or where the report will be read by a group rather than an individual.

2.6.2 An abstract should be used at the beginning of this format.

2.6.3 If the report is longer than seven or eight pages, a *formal* report format should be used rather than an article format.

2.6.4 The article format should be considered as an expanded abstract of the formal report. Intricate details and explanations are omitted in an article report.

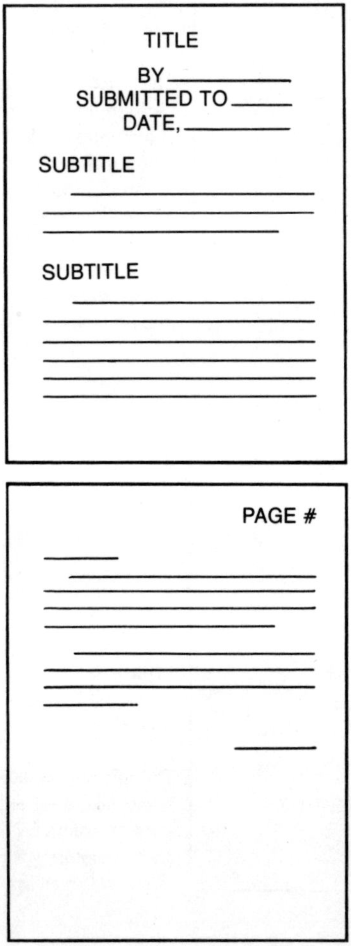

Figure 6 The Article Format

REPORT FORMATS

2.7 Formal Reports (see **Figures 7 - 12**)

The formal report necessitates the greatest amount of care in design. Each report presents a separate problem, and the characteristics of each problem require that certain sections be amplified and others reduced. The following example illustrates the basic components of the formal format:

Figure 7

Letter of Transmittal

The letter of transmittal should perform the initial introduction. It should present, in a clear and concise manner, the following items: authority for report, statement of problem, scope of investigation, limitations, principal conclusions and recommendations, methods used in investigations, names of co-workers, and acknowledgments.

Figure 8

Title Page

The title page should contain six main items: title of report, author, location (firm or institution), date, addressee, and laboratory (or course name and number), when applicable.

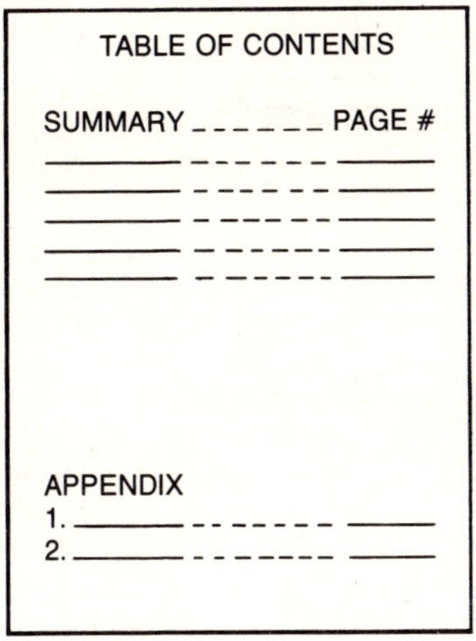

TECHNICAL REPORT STANDARDS

Figure 9

Table of Contents

All major divisions and important minor divisions should be presented by title and page number in the table of contents.

NOTE: It may be desirable to include a table of Graphs or Figures, if such information will prove useful to the reader.

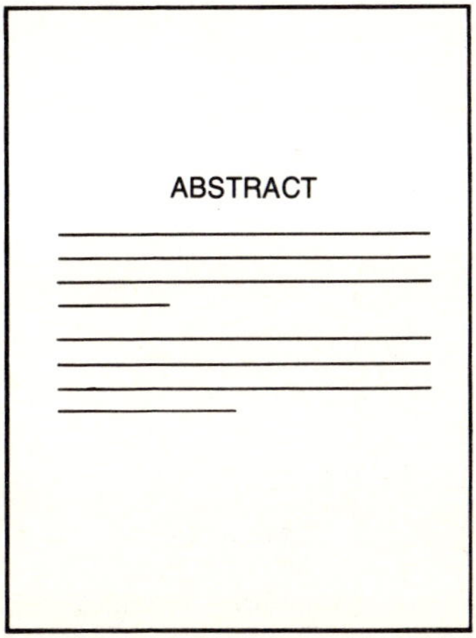

Figure 10

Abstract or Summary

The abstract should present a concise summary of the problem, findings, conclusions, and recommendations.

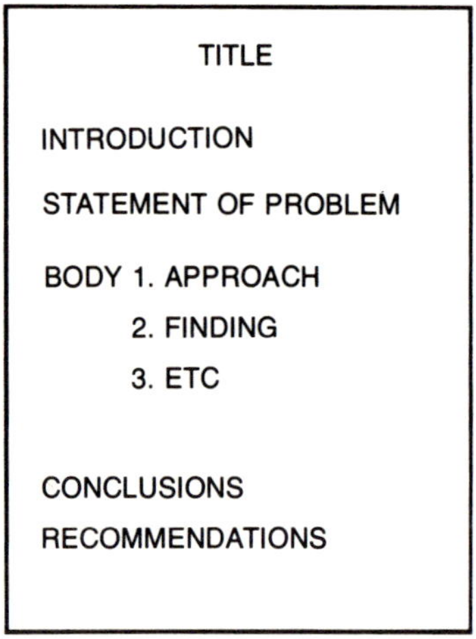

REPORT FORMATS

```
TITLE

INTRODUCTION

STATEMENT OF PROBLEM

BODY  1. APPROACH
       2. FINDING
       3. ETC

CONCLUSIONS
RECOMMENDATIONS
```

Figure 11

Report Body

A complete and comprehensive presentation of the material should comprise the main body of the formal report. An evaluation of the validity of the investigation should also be presented. The conclusions and recommendations should be tabulated if possible. In the body of the report, the procedures, observations, pertinent data, and sample calculations should also be presented.

```
APPENDIX
```

Figure 12

Appendix

The appendix should contain data sheets, graphs excluded from the text, bulky calculations, and other addenda. The references should precede the appendix.

SECTION THREE

TABLES

Tables are primarily used to present numerical or other briefly stated data and results. A table is preferred because of the order it brings to such groups of data and results. Moreoever, when numbers are placed in a table, errors are less likely to be made in the transposition of the figures for calculating, graphing, or other purposes.

It is important that all necessary information be included in the table. See **Figure 13**. Observe the following rules in the preparation of tables:

- **(A)** Every table should have a number and title. The numbering should be consecutive.
- **(B)** The title should state where the material was gathered and what it means.
- **(C)** All columns or rows which do not contain dimensionless numbers should include the units. This usually is done in the heading.
- **(D)** If possible, avoid abbreviations in the headings.
- **(E)** Tables should face the bottom of the page unless they accompany a graph which faces to the right. The reader should not have to rotate the report continually to read the tables.
- **(F)** Do not leave blank spaces in a table without explanations. Either write a zero in the space, if warranted, or indicate by a dash that no data were taken.
- **(G)** If the table is being used to present data, it is advisable to include the least reading and full scale of the instrument used. This information also may be presented in a footnote at the bottom of the table.

TABLES

(H) Never place a table on two pages if it will fit on one.

(I) Experimental values and calculated values should be recorded with no more significant figures than are warranted by the accuracy of the source.

Table 1. Properties of Gases (SI)

Gas	Symbol	Molecular Weight	$\gamma = \frac{c_p}{c_v}$	Gas Constant N – m/kg – °K R	Specific Heats J/kg – °K c_p	c_v	Viscosity N – s/m² μ
Air	—	28.97	1.40	287	1,000	716	1.8×10^{-5}
Argon	Ar	39.94	1.67	208	519	310	2.3×10^{-5}
Carbon Dioxide	CO_2	44.01	1.29	189	850	657	1.5×10^{-5}
Carbon Monoxide	CO	28.01	1.40	297	1,040	741	1.8×10^{-5}
Helium	He	4.00	1.67	2,080	5,230	3,140	2.0×10^{-5}
Hydrogen	H_2	2.02	1.41	4,120	14,300	10,200	9.1×10^{-6}
Methane	CH_4	16.04	1.32	519	2,230	1,690	1.1×10^{-5}
Nitrogen	N_2	28.02	1.40	296	1,040	741	1.7×10^{-5}
Oxygen	O_2	32.00	1.40	260	913	653	2.0×10^{-5}
Water Vapor	H_2O	18.02	1.33	461	1,860	1,400	1.1×10^{-5}

Tabular values are for normal room temperature and pressure.

Figure 13 Sample Table

SECTION FOUR
GRAPHS

4.1 Introduction

Unfortunately, standards in graphical communication are not uniformly accepted by research scientists, engineers, laboratory technicians and others engaged in technical report-writing. The standards which do gain acceptance generally are those promoted by the federal government through contract requirements or by professional societies and universities. These three sources have been the major influences for communication standardization.

With such a variable basis, it cannot be said that a specific graphical standard is correct or incorrect. It is possible, however, to present the preferred standards and methods prevalent in the technical professions today. The reader is advised to temper these rules and to extrapolate with good common sense when questions arise. Remember: the object is effective communication.

With this in mind, we present the following suggested and preferred guide to methodology in graphical communication.*

4.2 Planning the Graph

Often, the desired results of an experiment are graphs or illustrative comparisons. When not an end result, graphs should be used to supplement, complement, simplify or clarify the text of the report. If they do not do this, they probably should be omitted.

---

*In addition to the presentation of data in the Cartesian form covered in this section, there are a variety of other forms, such as bar graphs and pie charts. Some of these are discussed in *Principles of Experimentation and Measurement* by Gordon Bragg (Englewood Cliffs, N.J.: Prentice-Hall, Inc., 1974, Chapter 6).

GRAPHS

Consider the following factors when contemplating the use of graphs.

4.2.1 The nature of the individuals who are to be reached.

4.2.2 The critical or significant facts to be communicated.

4.2.3 The type of illustration or graph that will best communicate these facts.

4.2.4 The necessity or advisability of using a graph. Remember: graphs should not be used merely to add sophistication or fill space in a report.

The graphs must support the general purpose of the communication. Since they also serve to save the reader's time, the graphs should be simple and not difficult to read and interpret. They also should be free of material that is not essential to the reader's understanding.

4.3 Types of Relationships

It is important to understand the nature of the relationships that you will be presenting graphically. The three most common relationships are: observed, empirical, and theoretical. The appearance of the graph should indicate the type of the relationship.

4.3.1 Observed relationships

Observed or measured values, represented by points on a graph, should be connected with straight lines except when the general or specific shape of the curve is well known. **Figure 14** provides an illustrative example. Stress-strain

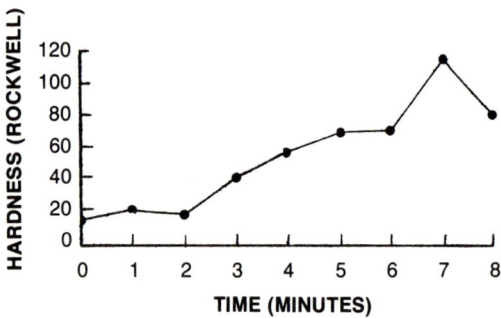

Figure 14 A Typical Observed Relationship

curves and transistor characteristic curves are examples of well-known curves. Data on these graphs may be connected with a smooth curve.

4.3.2 Empirical relationships

The empirical relationship generally denotes the author's interpretation of the observations. This is frequently referred to as a "least squares" or "best fit" method. It can be done mathematically or visually. The mathematical fitting usually is employed in a linear or almost linear relationship. Remember that the fitting of a curve to observed data should be reserved for definable functions, continuous functions, and functions whose variables can be readily related. **Figure 15** shows how this type of relationship should be plotted as well as the experimental data points.

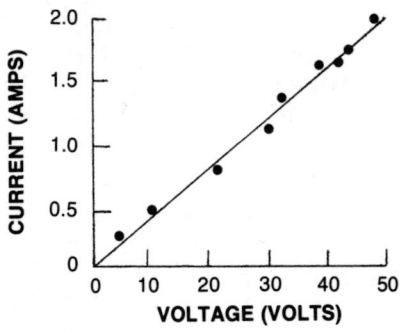

Figure 15 A Typical Empirical Relationship

4.3.3 Theoretical relationships

The theoretical relationship is plotted from known functions. The parameters and coefficients can be determined from the data, but the functions are predetermined. An example of this would be the relation $y = ae^{bx}$. Normally, **x** is interpreted as the independent variable, **y** as the dependent variable, and **a** and **b** as the coefficients, constants, or parameters. The plotted curve must be continuous if the function is continuous. Observed data may accompany a theoretical curve to show experimental deviations, as in **Figure 16**.

GRAPHS

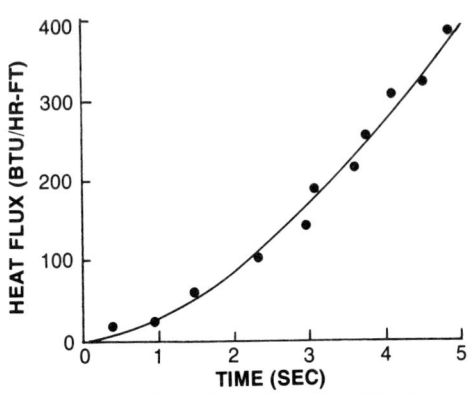

Figure 16 A Typical Theoretical Relationship
(Points are shown to illustrate discrepancy)

4.4 Variables

Generally, the independent variable of a relation is plotted on the *abscissa* (horizontal axis), and the dependent variable is plotted on the *ordinate* (vertical axis). When generally accepted standards exist which are contrary to this rule, however, these standards should be followed. Typical of the latter case are the conventional stress-strain curves.

The most common independent variables are time, distance, voltage, and load. For proper graphing, it is important to determine which of the variables is independent. The independent variable usually is the one which is controlled in an experiment; the dependent variable is the one which is observed or measured. **Figure 17** illustrates a typical **graph** consistent with the rule. **Figure 18** shows a common exception.

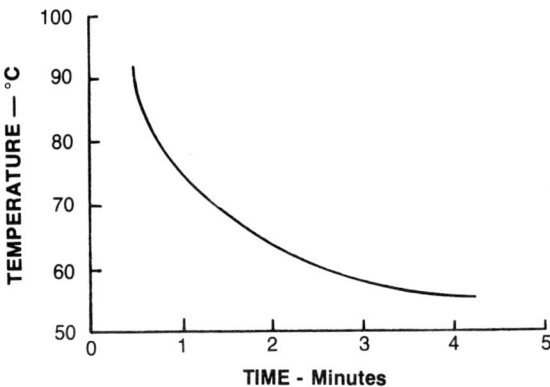

Figure 17 A Graph Which Follows The Independent - Dependent Variable Rule

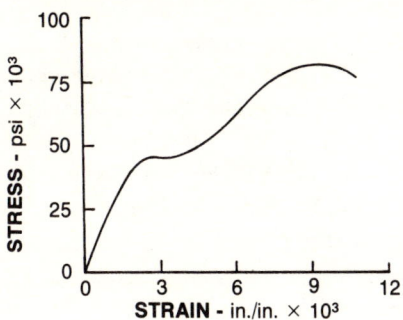

Figure 18 An Accepted Standard Which Deviates From The Rule

4.5 Size

The size of the graph should be determined on the bases of effectiveness and uniformity. Very large or very small graphs reduce the effectiveness of the communication. The simpler the graphic material, the smaller the graph can be made. Conversely, more complex graphs should be made larger.

Another consideration is the interpretive accuracy required. High accuracy demands a larger graph. Note that, contrary to popular opinion, finer subdivisions do not have this effect.

Probably the most important consideration is consistency. For example, if graphs are designed to obtain the stress-strain curves for several metals, then the graphs should be approximately the same size. One reason for maintaining size consistency is that it provides a systematic control of accuracy. This is particularly useful when graphical integration is performed. Also, uniform size facilitates comparisons among graphs.

4.6 Titles

The graph title must complete the story told by the graph. Explanatory material may be added as a subtitle if it provides needed clarity. The reader should not have to page through the report to determine the object to which the data refer. Thus, remember this rule: a graph must be correlated with the text and cited in the text. Too, the title should not take the form, "INDEPENDENT VARIABLE VERSUS DEPENDENT VARIABLE" or "STRESS VS. STRAIN." An adequate title also should cite the material tested as well as the type of test performed. For example, "TENSILE TEST OF 6061-T6 ALUMINUM." In particular, the graph title must not simply restate the axis titles.

Titles must be logically clear and concise. Abbreviations generally should be avoided. When in doubt, it is best to write out the words in full.

GRAPHS

The title should face either the bottom of the page or to the right, depending on which way the graph faces. The preferred standard is to have the graph and title face the same direction. In no event should the title or graph face to the left or to the top of the page.

The title should not be placed within the rectangular region of the graph unless the entire graph fills the page. The preferred position of the title is beneath the bottom of the graph.

Titles may be fully capitalized, or just the first letter of each word in the title may be capitalized. Never use all lower case. Whichever scheme is chosen, use it consistently throughout the report. Some examples are:

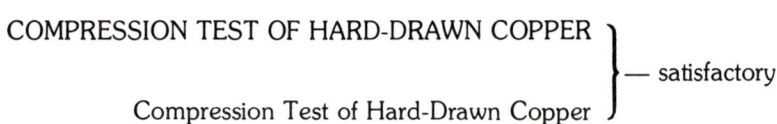

compression test of hard-drawn copper —unsatisfactory

When assigning numbers to graphs, keep them as simple as possible. A straight numbering system usually is preferable to one that involves decimals and large or small letters. Overly complex systems tend to defeat their own purpose.

The words "graph" or "figure" should be used, as follows:

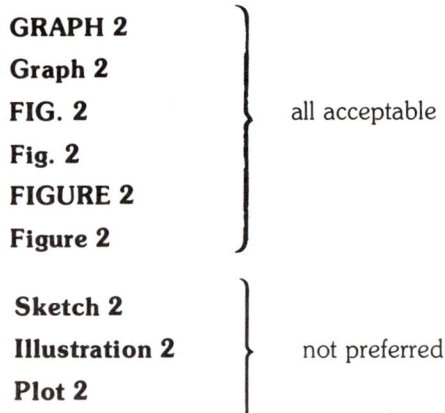

TECHNICAL REPORT STANDARDS

4.7 Axis Labels

Closely allied with standards in titles are the methods of labeling axes. Abbreviations should be avoided. While not incorrect, they are often confusing. For example, "Efficiency" should be used instead of "Eff". Similarly, do not use symbols unless they are accompanied by identifying descriptions.

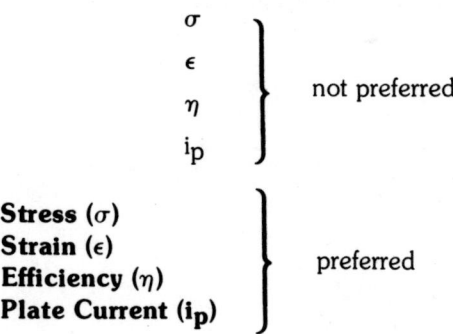

Either parentheses or a comma may set off the symbol. Therefore, both Stress, σ, and Stress (σ) are satisfactory. Parentheses or quotation marks should never enclose the entire axis title.

The labels of the axes must read from the right or bottom as the graph is observed. **Figure 19** illustrates the proper placement of axis labels.

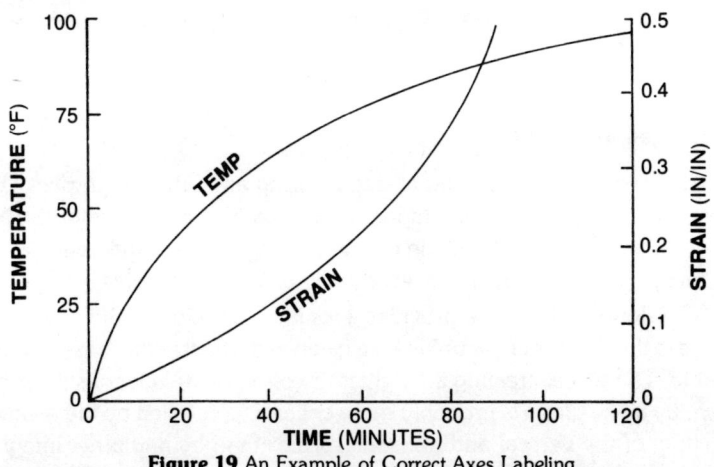

Figure 19 An Example of Correct Axes Labeling

GRAPHS

Ordinarily, the ordinate should not be labeled perpendicularly to the axis line. The ordinate label should never face the left of the graph as it is read.

The titles of the axes should be in the same size writing or printing. The size used should not exceed the size of the title of the graph.

If magnitudes are not shown and all that is desired is a general shape, then the axes may be labeled as shown in **Figure 20.** In all graphs where magnitudes are shown, omit the arrows.

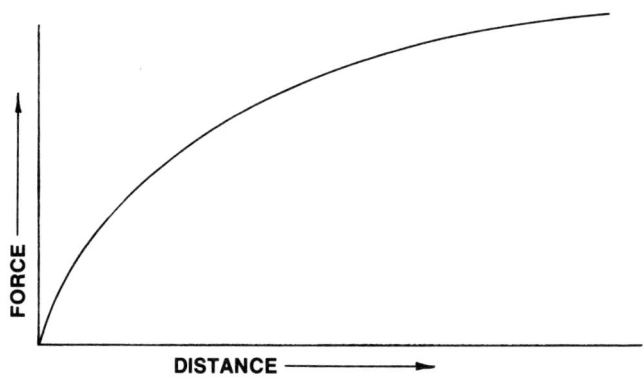

Figure 20 A Generalized Curve

4.8 Scales and Units

Scale selection is an integral step in sizing and labeling graphs. The choice of scales is extremely important and cannot be overemphasized. Identical data can be presented in two graphs, each having different scales, and these two graphs may convey entirely different messages.

The slope of the curve provides a visual impression of the degree of change in the dependent variable for a given increment in the independent variable. Therefore, creating the right impression of the relationship to be shown by a line graph is probably more critically controlled by the relative stretching of the vertical and horizontal scales than by any other integral feature. We have attempted to illustrate this effect in **Figure 21.**

TECHNICAL REPORT STANDARDS

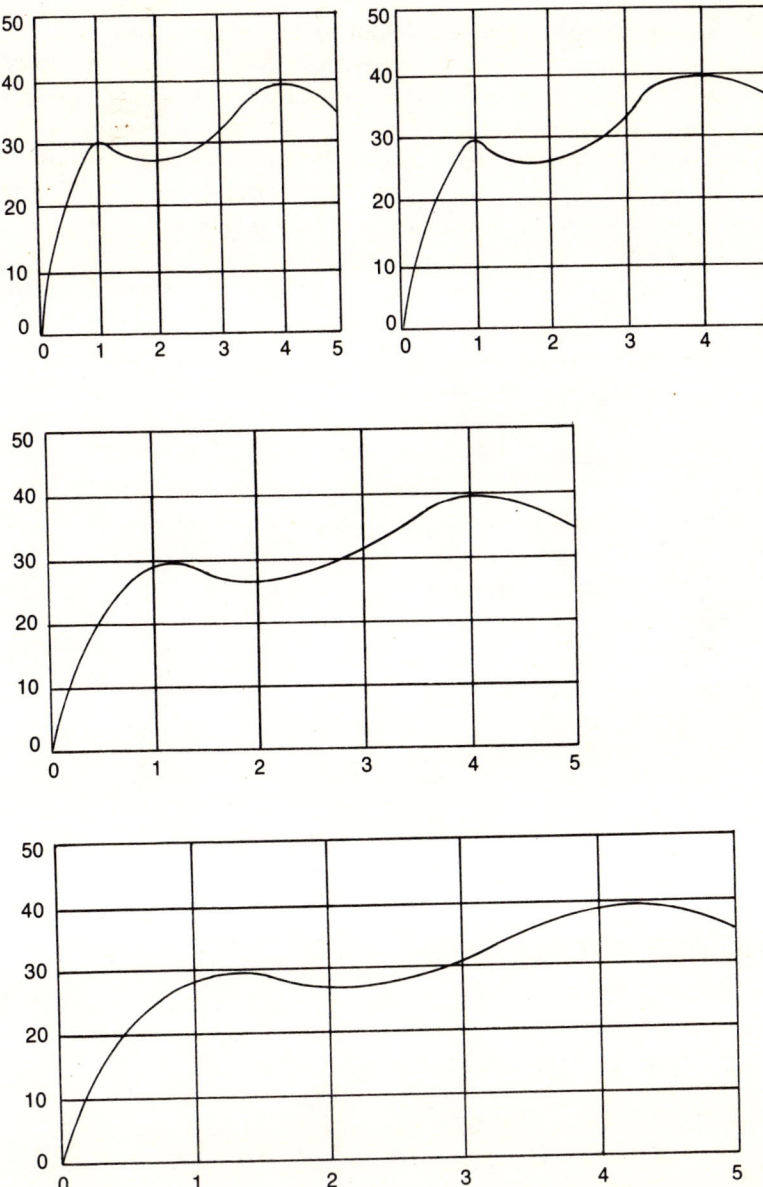

Figure 21 The Effect of Choice of Scale

GRAPHS

By expanding or contracting the vertical scale relative to the horizontal scale, a given difference in magnitude of the dependent variable can be made to appear of great or little significance with regard to (a) physical magnitude, (b) sensation, or (c) value, depending on the author's purpose. A large angle of slope, for example, of over 30 or 40 degrees, ordinarily is interpreted psychologically as of great significance. Conversely, a small angle, say of less than five degrees, usually is interpreted as of little significance.*

Sometimes complete freedom of scale selection is not possible. This is the case when using a given grid or coordinate system. Nevertheless, the ideal size should not be sought at the expense of a poor scale.

All regular graphs should normally have a zero reference. A zero point with a broken line may be used in combination when the relationship to zero is not significant. In some cases, the zero line or point may be entirely omitted when its presence reduces the effectiveness of the communication or is totally unrelated to the message. Such omission is at the discretion of the author.

Two or more scales should not be used on the same axis. This tends to reduce the clarity of the graph.

The digits used to label axes should be kept to a minimum and written without ambiguity. Consider the following general examples:

Poor Form	**Good Form**
1.0000	1
.12	0.12
100,000	10×10^4

---

*American Standard-Engineering and Scientific Graphs for Publications, American Society of Mechanical Engineers, 1947.

TECHNICAL REPORT STANDARDS

As indicated in these examples, large or small axis values should be stated in scientific notation for brevity. There are three common ways of expressing such quantities within the axis title. For example, let us say an axis scale ranging from zero to 100,000 BTU is numbered from zero to ten. The three alternative ways of stating these values in the axis title are: "Thermal Energy $\times$ 10^4 BTU," "Thermal Energy, BTU $\times$ 10^{-4}," and "Thermal Energy, ten-thousands of BTU's."

The axes of all graphs must end on a scale increment. This rule is illustrated in **Figure 22.**

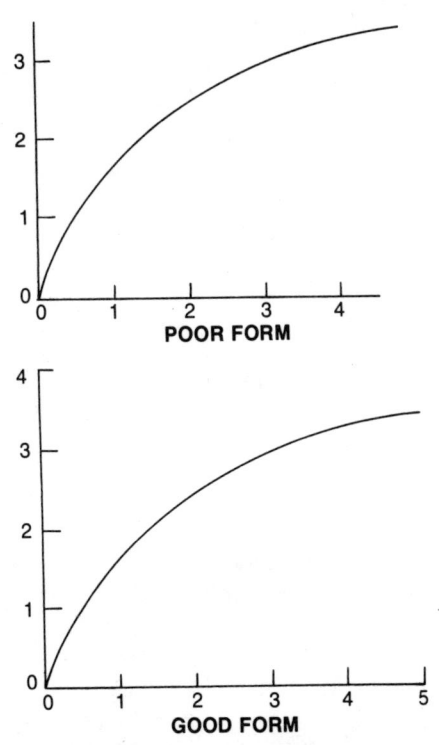

Figure 22 The Labeling of Axes

Another important rule, illustrated in **Figure 23,** is: be careful not to divide an axis into too many increments. The result is clear in the figure.

GRAPHS

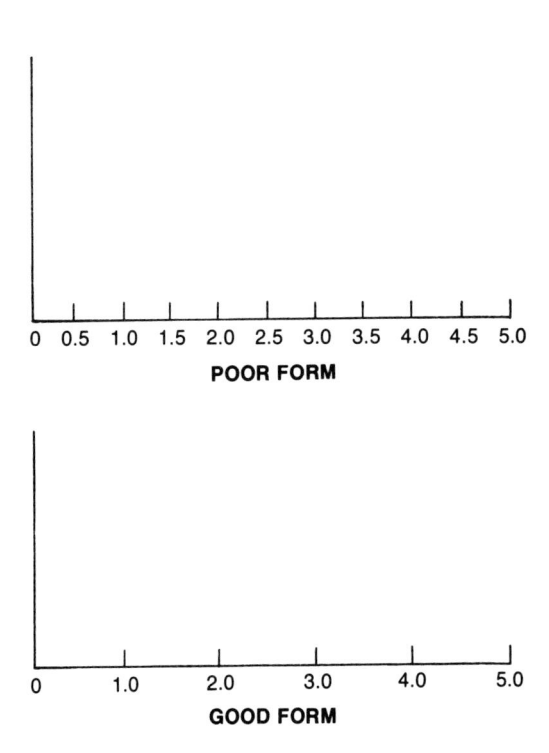

Figure 23 Scale Choices

A good general rule to follow in establishing scale values is that the smallest division of a graph should represent one, two, or five basic units, or multiples of ten thereof. This means that, on ruled graph paper, the smallest ruled division, scale unit or not, should be determined by this guideline. On unlined paper, the indicated scale units also must adhere to this rule. In either case, the numbers used to label axes (scale units) should also be whole integers, one-half integers, or one-fifth itegers.

An important standard to remember in drawing graphs is that the axis lines and scales always must cover at least the region of the curve. In other words, the curve and/or data points should never extend beyond the indicated axes.

TECHNICAL REPORT STANDARDS

4.9 Symbols

If experimental data are plotted, the data points must be shown. Preferably, the associated curves should be represented by solid lines. When more than one curve is to be presented on a graph, the curves can be distinguished by using different types of lines, different widths, or characteristic symbols. Labeling also can be used. Symbols are utilized to designate observed points. Calculated points also can be set off with symbols when the function is complex or discontinuous, or the shape is uncommon or unfamiliar.

Solid and dashed lines also are suitable for designation when two or three curves are shown close together. When comparing calculated results to experimental results, solid and dashed lines are often a good means of differentiation, as illustrated in **Figure 24**.

When symbols are used, they should be circles, squares, or triangles (○, □, ∆) rather than solid marks or X's (X, +, ●, ■, ▲). Though not a rigid standard, this is recommended to prevent the symbol from blotting out the data point.

Curves also may be designated by brief labels placed close to them (either horizontally or along the curve).

Keys should be used with symbols. Place the key within the coordinate system and set off with a border.

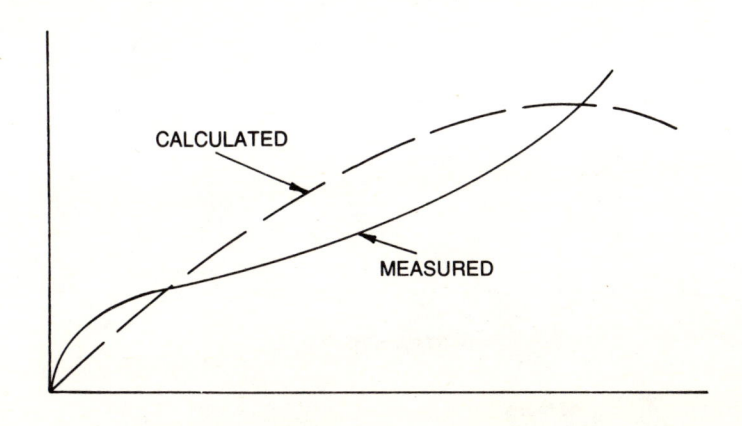

Figure 24 Designating Curves

4.10 Information on Graphs

If plotting a given curve, be sure to state the reference in the title, subtitle, or on the graph. The basis of all data must be clear. Supplementary information can be placed near the graph but should not be placed on the graph unless absolutely necessary. In any event, the sources used in plotting should be indicated.

4.11 Questionable Points and Deletions

An erroneous plot (point) should be deleted with an "X." If a range of values is deemed questionable, indicate this in a suitable manner on the graph as well as in the text. Some suggestions for this are presented in **Figure 25.**

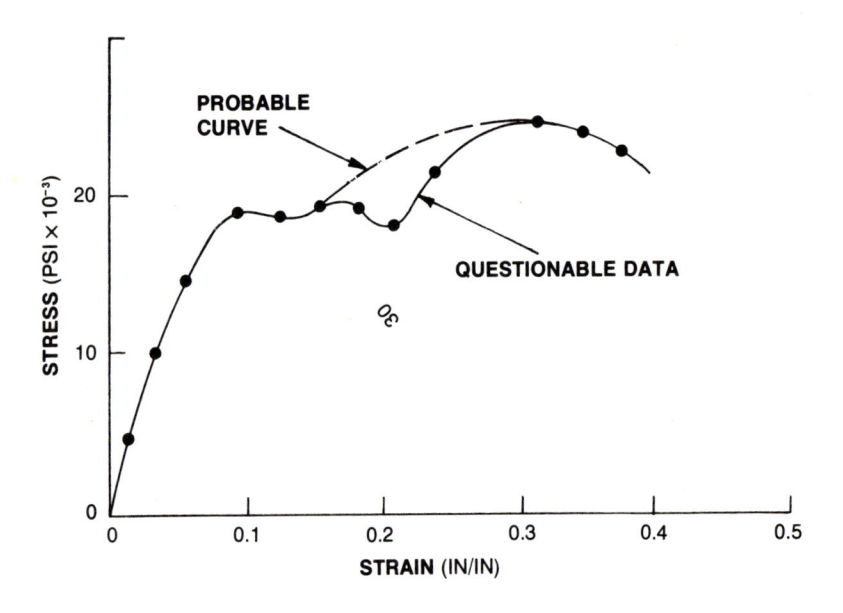

Figure 25 Illustration of Questionable Data

4.12 Folding Large Graphs

If a finished graph is too large to fit in the text of a report, proper folding is essential so that the graph can be unfolded easily. A sample is shown in **Figure 26.** Make sure that the title is visible without unfolding the graph.

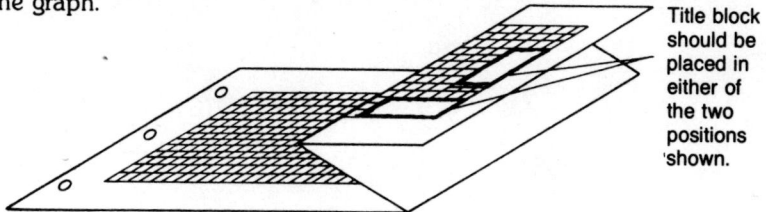

Title block should be placed in either of the two positions shown.

Figure 26 The Proper Folding of a Large Graph

SECTION FIVE

ILLUSTRATIONS

The basic rules of good graphics also apply to illustrations and figures. It is up to the author to determine what materials will be most effective if illustrations are included. The three main types of illustrations are: graphs, circuit diagrams, and drawings. The rules governing graphs were set forth in the preceding section. Some basic guides to the use of circuit diagrams and drawings will be presented here.

Illustrative material which is used to support the text of a report should be well integrated into the body of the report. If illustrations are presented for general information, and are not essential to the clear understanding of the text, then such illustrations should be placed in the appendix or at the end of the report. When illustrations are used in the text, place them as close as possible to the related discussion. In no event should illustrations precede the information they support. If the illustrations used are large (eight inches or more in height), then a separate page may be used in the text. If the drawings are not only large but also complex, then it is preferable to place them in the appendix and merely refer to them in the text. Remember: too many graphs, drawings, and diagrams in the body of a report tend to break up the flow of the presentation.

Neither circuit diagrams nor drawings should be placed so that they face the right of a page unless absolutely necessary, for this necessitates turning the report while reading it. It is your responsibility to save the reader's time. The same rule of good placement also apply to labels, dimensions, and titles. Make sure your writing can be read from the same position that the illustration is viewed. Normally, both circuit diagrams and drawings can be labeled so they can be read from one direction, a single orientation.

ILLUSTRATIONS

The rules of drafting are also important. A basic knowledge of drafting standards is prerequisite to preparing good illustrations. Consequently, it is expected that the author will adhere to these standards, particularly in designing circuit diagrams. Similarly, a compass and straight edge should be used when preparing drawings. Free-hand sketches are permitted in notebooks only if they are neat and carefully drawn.

All drawings and diagrams must be properly labeled with a number and title. This title, including the figure number, should be referred to in the text. If it is not, then the illustration probably is not important and should be omitted or placed in the appendix.

Colors should not be used in illustrations. Line variations easily provide contrast. Other means of distinguishing lines have been discussed in the previous section. This non-color rule is particularly important since reports are frequently photocopied, and most copying processes do not reproduce all colors. Thus, pertinent information could easily be lost.

SECTION SIX

ERRORS

6.1 Introduction

When a report is written about laboratory work or findings in which measurements have been made, it should contain an error discussion. This need not be a written discussion, but the report must present some evidence of an analysis of the possible errors involved. Error analysis can be expressed either in tabular or graphical form. Perhaps the most common means is by showing range markers on the graphs of the experimental results.

Always include an error discussion in any laboratory work or findings in which measurements have been made. This need not be a written discussion, but the report must present some evidence of an analysis of the possible errors involved. Error analysis can be expressed either in tabular or graphical form. Perhaps the most common means is by showing range markers on the graphs of the experimental results.

In any event, the report should contain sufficient information so that the reader may easily read or see the accuracy of the results. The value of a laboratory or research report is virtually nil if it makes no effort to present and analyze the errors encountered.

The inclusion of an error discussion or evaluation fulfills two purposes. First, it provides an immediate figure of merit for the reader. Second, it also may provide clues to another researcher, as well as to the author, as to where changes might be made to improve or extend the work.

ERRORS

6.2 Definition of Terms

In this section, we will define several common types of errors and related quantities. These definitions are fairly common in statistical error analysis, but they also have a definite place in the type of analysis presented here.

6.2.1 Errors

An "error" is defined as the difference between a measured value and the "true" or accepted value. In many cases, the "true" or accepted value is not known, and we must rely on *estimated errors* or *discrepancies*. This section illustrates one way to make an analysis of such errors. Errors are commonly expressed as ± their magnitude and thereby carry the units of the variable measured. Errors also are expressed, at times, as percentages. Such values are obtained by dividing the error by the "true" or accepted value, if it is known. If not, the percentage error may be estimated by dividing the error by the measured value.

When accumulating errors from a series of terms in an expression or calculation, the errors are always considered positive and are added together to obtain the largest possible estimate of the error. The plus-or-minus sign is employed only to indicate that it is not known where the actual or "true" value lies. If M denotes a measured value and ΔE the experimental error, then the estimated value is denoted $M \pm \Delta E$. This expression indicates that the "true" value of the measured quantity is anticipated to be somewhere in the region $M - \Delta E$ to $M + \Delta E$.

6.2.2 Random errors

Whenever a series of measurements of a quantity has been taken, one generally expects the errors or discrepancies to occur randomly at each measurement. Random errors are caused by not reading an instrument scale from the same position each time, by friction in the meter movement, fluctuations in supply voltage, and so forth. To illustrate this and other types of errors, we shall utilize a series of bull's-eye targets. Random errors are illustrated in **Figure 27**.

TECHNICAL REPORT STANDARDS

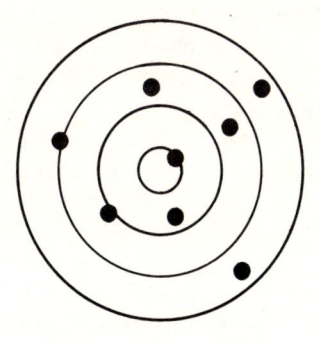

Figure 27 Random Error

6.2.3 Systematic errors

When the errors in a series of measurements exhibit a constant bias or deviation from the expected value, the error is said to include a "systematic" error. If such an error is discovered, it may be corrected by subtracting its estimated value from each measurement in which it occurs. The bull's eye target in **Figure 28** illustrates the occurrence of systematic errors.

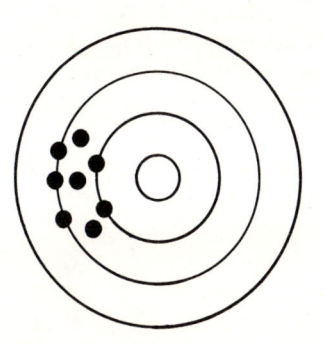

Figure 28 Systematic Error

Note that the entire group is displaced from the bull's eye by a constant factor. Several sources of systematic errors are: personal errors, such as reading a meter scale from a constant view point which is not in proper alignment; calibration error in the instrument; a bent pointer or indicator; incorrect zero adjustment; and damaged sensing elements or shunt resistors.

6.2.4 Precision

A series of measurements is said to be "precise" if the random errors are small. "Precision" is a term used to denote the relative magnitude of the random errors irrespective of any systematic errors. In terms of our bull's-eye, "precision" refers to the tightness or closeness of the group.

6.2.5 Accuracy

A series of measurements is said to be "accurate" if the systematic errors are small. In contrast to precision, "accuracy" is used to denote the relative magnitude of any systematic errors. **Figure 29** illustrates various combinations of accuracy and precision. Note that the concepts of "accuracy" and "precision" are not clearly distinct.

6.2.6 Blunders

Any introduction of error into an experiment from misinterpretation of scale readings, incorrect determination of scale factors, and other misjudgments is termed a "blunder." This type of error can be prevented by following careful procedures and knowing what is taking place. It should be noted that this type of error is not acceptable in presenting experimental results.

6.2.7 Computation Errors

Computation errors are those which are introduced into experimental calculations through the devices used to perform the calculations, such as a slide rule or calculator. Once the experimental data have been recorded, careful selection of the equipment to perform the calculations is very important. Computational errors may be kept to a minimum if the errors of the selected equipment are insignificant with respect to the experimental errors. For example, if the experimental data are recorded to five or six significant figures, then the use of a slide rule for computation would introduce computational errors which are far greater than the experimental error. It must be remembered that mathematical mistakes are blunders, not computation errors.

TECHNICAL REPORT STANDARDS

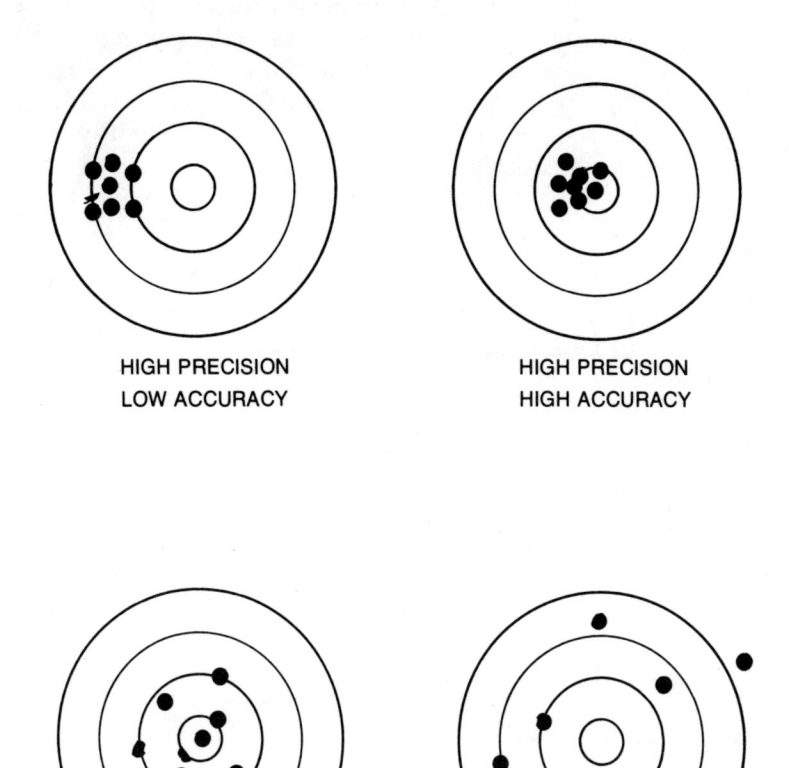

Figure 29 Examples of Accuracy and Precision

ERRORS

6.2.8 Instrument precision

The "precision" of an instrument is defined as plus or minus one-half the smallest division (i.e., least reading) of the scale. As an example, consider the scale shown in **Figure 30**. For this intrument, the indicated precision is ± 0.1 volt dc. The precision expresses the closeness to which the scale may be reliably read.

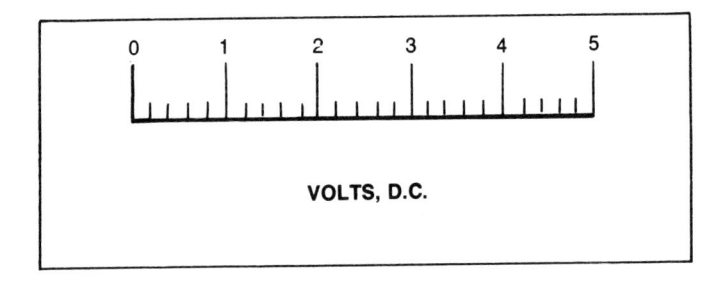

Figure 30 Sample Meter Scale

6.2.9 Instrument accuracy

The "accuracy" of any instrument is defined as the ratio of one-half the least reading to the full scale reading. It usually is expressed as a percentage. For the scale of **Figure 30**, the accuracy is $\pm \frac{1}{2}$ ($\frac{0.2}{5}$) = ± 0.02 or ± 2.0 %. It should be noted that the accuracy is based on the full scale reading; thus, it is equal to the error in reading a voltage equal to the full scale voltage only. The error in reading any voltage less than the full scale value is ± 0.1 volt. The percentage error in reading values less than the full scale value is the ratio of the full scale value to the actual reading times the instrument accuracy. If, for our sample meter, we had a reading of 3.4 volts, then the accuracy would be $\pm \frac{5.0}{3.4} \times 2.0\% = \pm 2.9\%$. An alternative way of estimating the error in reading the value 3.4 volts is to divide the precision by the value read: $\pm \frac{0.1}{3.4}$ = ± 0.029 or ± 2.9%.

At this point, one might justifiably question the validity of defining instrument accuracy and precision in terms of the scale markings rather than according to the manufacturer's values for the meter movement. The fact is that, no matter how accurate the movement is, the meter can be read with no greater accuracy than that of the scale markings. No instrument maker would consider placing a scale on a meter which was much less accurate than the movement itself.

On the other hand, a manufacturing firm which used a scale much more accurate than its movement would soon lose all of its customers. This would be particularly true if it advertised its instruments as having the accuracy indicated by the scales. Clearly, the common practice among manufacturers is to match the scale accuracy to the movement as closely as possible.

It should also be noted that the percentage error in reading a linear scale varies inversely with the reading. However, instruments with logarithmic scales maintain a constant percentage error in reading scale values.

6.2.10 Instrument sensitivity

The "sensitivity" of an instrument is its degree of response to stimuli. It is the ratio of the indicated change of deflection to a change in the measured quantity. More commonly, the sensitivity is expressed as the total resistance divided by the voltage required to produce a full scale deflection (i.e., ohms per volt). Defined in this manner, "sensitivity" is the current required by the meter movement to produce a full scale deflection.

6.3 Error Evaluation

The actual evaluation of errors involved in any experiment can become difficult depending upon the extent of the desired evaluation. Three relatively easy methods of error evaluation are presented and explained in this section. These methods should be used when only a few readings are taken for each specific test point or setting. When a larger number of readings are taken, these methods may be applied to the mean value, but it is recommended that the more common and useful method of

ERRORS

statistical analysis be employed.

While methods of statistical error analysis are not presented in this book, they are available to the reader in any standard text on statistics. Two particularly good references on statistical error analysis are those by Beers and Bragg (see SUGGESTED REFERENCES).

In the previous discussion, the words "few" and "a larger number" used with regard to readings are quite ambiguous because it is difficult to cite how many readings comprise this dividing point. As an approximation, however, we will estimate the figure of 10 as a useful dividing line for the application of either of the two methods. The specific choice is up to the investigator and depends upon the demands of the experiment undertaken.

6.3.1 Algebraic relations

The formulas presented in this section are based on the assumption that a "true" value, say A, is composed of the measured value, A', and the error or uncertainty, ΔA. Specifically, this is written as follows:

$$A = A' \pm \Delta A$$

In estimating the error for a quantity which is determined by an algebraic relation between two or more other quantities, it can be assumed that these other quantities are represented in the same manner as for A. Using various means of expansions, we determine a formula for the estimated error in the desired quantity. In reaching this point, two basic assumptions are followed:

(a) It is assumed that the uncertainties in any measurement are small enough that the products and squares as well as higher order terms involving the uncertainties may be neglected.

(b) To obtain the largest possible estimate of the total uncertainty, the signs of the partial uncertainties are always neglected or considered positive so that all the terms are added together.

In an actual experimental situation, the measured values are the readings taken from the measuring devices. The uncertainties derive from

TECHNICAL REPORT STANDARDS

the precision or accuracy of the instruments used. As noted earlier, an instrument's precision equals the accuracy times the full scale reading. As an example, assume that voltage measurements are being made with a meter whose scale is depicted in **Figure 30**. The uncertainty to be used for all the measurements would be the instrument's precision, or ± 0.1 volt.

To illustrate the procedures used in estimating errors and to provide examples of the methods employed, the following sample problems will be considered. First, assume that we have three known or measured quantities:

$$A = 4.00 \pm 0.20 = A' \pm \Delta A$$
$$B = 7.00 \pm 1.00 = B' \pm \Delta B$$
$$C = 5.00 \pm 0.50 = C' \pm \Delta C$$

In the first example, we will determine the error or uncertainty in the quantity

$$X = A + B - C$$

The problem is to determine the terms X' and ΔX for the relation

$$X = X' \pm \Delta X$$

By substituting into the given equation, the result is

$$X = (A' \pm \Delta A) + (B' \pm \Delta B) - (C' \pm \Delta C)$$
$$= A' + B' - C' \pm (\Delta A + \Delta B + \Delta C)$$

The last term was obtained by employing assumption **b** above. From this result, it is seen that

$$X' = A' + B' - C' = 4.00 + 7.00 - 5.00 = 6.00$$
$$\Delta X = \Delta A + \Delta B + \Delta C = 0.20 + 1.00 + 0.50$$
$$= 1.70$$

hence,

$$X = 6.00 \pm 1.70$$

ERRORS

If the result is expressed in terms of percentage error, the calculation would be

$$X = 6.00 \pm \left(\frac{1.70}{6.00}\right) \times 100 = 6.00 \pm 28.3\%$$

To evaluate the estimated error for a product, consider the relation

$$Y = AB$$

Substituting the general expressions, the result becomes

$$Y = (A' \pm \Delta A)(B' \pm \Delta B)$$

Expanding the product and applying assumption **a**, the result is

$$Y = A'B' \pm (A' \Delta B + B' \Delta A)$$

Now, from this expression, it can be seen that

$$Y' = A'B' \, , \, \Delta Y = A' \Delta B + B' \Delta A$$

A more convenient expression for the estimated error may be obtained by dividing the latter expression by the equation for Y and applying assumption **a**, with the result that $AB \cong A'B'$

$$\frac{\Delta Y}{Y} = \frac{\Delta A}{A'} + \frac{\Delta B}{B'}$$

The terms $\Delta A/A'$ and $\Delta B/B'$ are the percentage errors in A and B respectively. Thus, the preceding expression states that, for a product, the percentage error is the sum of the percentage errors of the two terms.

Using the numerical values, the following is obtained

$$\frac{\Delta Y}{Y} = \frac{0.20}{4.00} + \frac{1.00}{7.00} = 19.3\%$$

As a final result,

$$Y = 28.0 \pm 19.3\%$$

TECHNICAL REPORT STANDARDS

A more common way of obtaining the same result is by using logarithms and taking total differentials. As an example, consider the previous problem. Taking the logarithm of the original equation gives

$$\log Y = \log A + \log B$$

The total differential of this expression is

$$\frac{\Delta Y}{Y} = \frac{\Delta A}{A} + \frac{\Delta B}{B}$$

Since the "true" values of A and B are not known, they can be replaced by their measured values A' and B' to obtain

$$\frac{\Delta Y}{Y} = \frac{\Delta A}{A'} + \frac{\Delta B}{B'}$$

This procedure may be verified by applying assumption **a**. Note that this procedure leads to the same result with greater ease.

As a further example of this method, consider

$$Z = AB/C = Z' \pm \Delta Z$$

Again, using logarithms, we find

$$\log Z = \log A + \log B - \log C$$

Taking the total differential and utilizing assumption **b** gives

$$\frac{\Delta Z}{Z} = \frac{\Delta A}{A} + \frac{\Delta B}{B} + \frac{\Delta C}{C}$$

After substituting the given values * and performing the calculations, the final result is

$$Z = 5.60 \pm 29.3\%$$

---

*Rather than repeat the discussion of replacing A, B, and C with the measured values A', B', and C', this operation should be followed automatically in all such computations.

ERRORS

To illustrate how an error analysis is made for a relation involving sums and products, consider the following:

$$W = \frac{A}{B} + C = W' \pm \Delta W$$

To analyze this expression, each term must be taken separately. For the first term, the result is

$$W_1 = \frac{A}{B} = W'_1 \pm \Delta W_1$$

Now,

$$\log W_1 = \log A - \log B$$

Proceeding as in the previous example, it can be found that

$$\frac{\Delta W_1}{W_1} = \frac{\Delta A}{A'} + \frac{\Delta B}{B'}$$

or

$$\Delta W_1 = W_1 \left(\frac{\Delta A}{A'} + \frac{\Delta B}{B'} \right)$$

Now, combine the terms to evaluate the total uncertainty.

$$W = \frac{A}{B} + C = W_1 + C = W' \pm \Delta W_1 + C + \Delta C$$

Thus,

$$W' = W_1' + C' = \frac{A'}{B'} + C' = \frac{4.00}{7.00} + 5.00$$

$$= 5.57$$

and

$$\Delta W = \Delta W_1 + \Delta C$$

$$= \left(\frac{A'}{B'} \right) \left(\frac{\Delta A}{A'} \pm \frac{\Delta B}{B'} \right) = \Delta C$$

$$= \left(\frac{4.00}{7.00} \right) \left(\frac{0.20}{4.00} + \frac{1.00}{7.00} \right) + 0.50$$

$$= 0.61$$

TECHNICAL REPORT STANDARDS

The final result is

$$W = 5.57 \pm 0.61$$

To further illustrate the method, the following five sample problems are provided:

$$R = (A + 3B)^{1/2} = R' \pm \Delta R$$
$$S = C^{1/3} = S' \pm \Delta S$$
$$T = A + B^2 - C = T' \pm \Delta T$$
$$U = \frac{AB}{C^{1/2}} = U' \pm \Delta U$$
$$V = \left(\frac{A}{B}\right)^3 + C = V' \pm \Delta V$$

Using the stated values,

$$R' = [4.00 + (3)(7.00)]^{1/2} = 5.00$$
$$S' = (5.00)^{1/3} = 1.71$$
$$T' = 4.00 + (7.00)^2 - 5.00 = 48.0$$
$$U' = \frac{(4.00)(7.00)}{(5.00)^{1/2}} = 12.5$$
$$V' = \left(\frac{4.00}{7.00}\right)^3 + 5.00 = 5.19$$

and

$$\Delta R = (1/2)[0.20 + (3)(1.00)] = 1.60$$
$$\Delta S = (1/3)(0.50) = 0.17$$
$$\Delta T = 0.20 + (2)(1.00) + 0.50 = 2.70$$
$$\frac{\Delta U}{U} = \frac{0.20}{4.00} + \frac{1.00}{7.00} + (1/2)\frac{0.50}{5.00} = 24.3\%$$
$$\Delta V = 3\left(\frac{4.00}{7.00}\right)^3\left(\frac{0.20}{4.00} + \frac{1.00}{7.00}\right) + \frac{0.50}{5.00} = 0.21$$

ERRORS

thus,

$$R = 5.00 \pm 1.60$$
$$S = 1.71 \pm 0.17$$
$$T = 48.0 \pm 2.7$$
$$U = 12.5 \pm 24.3\%$$
$$V = 5.19 \pm 0.21$$

6.3.2 Functional Relationships

Another type of relationship commonly encountered in experimental work is the *functional* one. This means that, instead of the type of algebraic relations previously discussed, the relation would be

$$A = \text{function of } x = f(x)$$

where x is the measured quantity. For example, f(x) could be a trigonometric or exponential function. Our problem now is as follows: having measured a value for x, that is, $x' \pm \Delta x$, determine the value of A and its estimated error,

$$\Delta A, \text{ that is, } A' \pm \Delta A.$$

To obtain $A = A' \pm \Delta A$ from $f(x) = f(x' \pm \Delta x)$,

we expand $f(x)$ in a Taylor's series about the point x'.

$$f(x) = f(x' \pm \Delta x) = f(x') \pm \frac{df}{dx}(x')\Delta x + \ldots$$

Here, ... represents terms of order $(\Delta x)'$ and higher.

If rule **a** is applied, the result is

$$A = A' \pm \Delta A = f(x') \pm \frac{df}{dx}(x')\Delta x$$

50

TECHNICAL REPORT STANDARDS

thus,

$$A' = f(x') \qquad \Delta A = \frac{df}{dx}(x')\,\Delta x$$

As an illustration, consider the function

$$A = \sin(x)$$

Let us also assume that the measured value of x is

$$x = 63° \pm 3° = x' \pm \Delta x$$

To correctly apply the relations just discussed Δx must be expressed in terms of radians:

$$\Delta x = \pm 3° = \pm 0.052 \text{ rad.}$$

Now, according to the given equation,

$$A = \sin(63°) \pm \cos(63°)(0.052)$$
$$= 0.89 \pm 0.02$$

For convenience, a few common functions are listed here with their approximate expansions.

$$\sin(x' \pm \Delta x') = \sin(x') \pm \cos(x')\,\Delta x$$
$$\cos(x' \pm \Delta x') = \cos(x') \pm \sin(x')\,\Delta x$$
$$e^{(x' \pm \Delta x)} = e^{x'} \pm e^{x'}\,\Delta x = (1 \pm \Delta x)e^{x'}$$
$$\ln(x' \pm \Delta x) = \ln(x') \pm \frac{\Delta x}{x'}$$
$$(1+x)^K = 1 + Kx \text{ provided } |x| < 1 \text{ and } K = \text{real}$$

6.4 Graphical Representation of Errors

At times, it may be advantageous to illustrate errors in data measurements directly on graphs in which the results are presented. The common manner in which this is accomplished is shown in **Figure 31**.

ERRORS

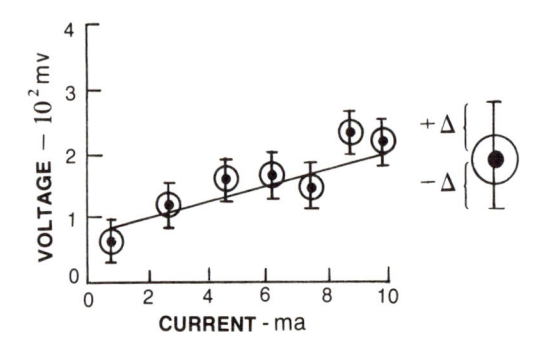

Figure 31 Graphical Representation of Errors

6.5 Significant Figures*

An important aspect of error analysis and mathematical computations is the consideration of significant figures. Calculations and other numerical results should not be carried out to more significant figures than known for the data. For example, one percent accuracy means that only three figures are significant. Four figures represent an accuracy of $\pm$ 0.1%, five $\pm$ 0.01%, and so forth. To illustrate this, all of the following represent numbers accurate to three significant figures: 507, 6.89, 0.00294, 4.25×10^{-6}. Note that, without further information, one would not know whether the number 800 had one, two, or three significant figures. The same would be true for 127,000. If this number were known to three significant figures, then it should be written 1.27×10^5. This illustrates another advantage of scientific notation.

In algebraic operations, the following rules are standard for considering significant figures:

6.5.1 Rounding off nonsignificant figures

When the value of the rejected digits is less than half a unit, the last digit retained is left unaltered. If the value of the rejected digits is greater than half a unit, the last digit retained is increased by one unit. When the value of the rejected digits is exactly half, the last digit retained is increased by one if it is odd and left unaltered if it is even.

---

*This material is in accordance with the standards of the International System of Units (SI), c.f. *ASTM Metric Practice Guide*, E380-72, Sec. 4.3 & 4.4, June 1972.

TECHNICAL REPORT STANDARDS

These rules are illustrated in the following examples, which are rounded to three significant figures:

$$14{,}238 \rightarrow 14{,}200$$
$$14{,}272 \rightarrow 14{,}300$$
$$14{,}250 \rightarrow 14{,}200$$
$$14{,}251 \rightarrow 14{,}300$$
$$14{,}372 \rightarrow 14{,}400$$

6.5.2 Addition and subtraction

When adding or subtracting, the answer may have no more significant digits than the least occurring in the sum. The proper procedure to follow in adding or subtracting numbers of varying significant digits is this: round all numbers to one place to the right of the least significant place and add, rounding the sum to one more place. The following example illustrates this procedure:

529.843	529.84
21.3	21.3
118.05	118.05
669.193 (incorrect)	669.19 = 669.2 (correct)

6.5.3 Multiplication and division

The result of multiplying and/or dividing a series of numbers must be rounded to no more significant figures than the least number in any quantity used. Thus, 21.0×134.101 is 2820, not 2816.1210.

6.5.4 Using logarithms and scientific notation

If logarithms are to be used for multiplying or dividing, a five-place table should be used if the numbers are known within 0.01 percent. A four-place table should be used when the numbers are known to 0.1 percent. If the data are no more than one percent accurate, a slide rule should be used.

ERRORS

In using scientific notation, powers of ten, the number is written as a product of two factors. The first part contains the significant figures and is always written with the decimal to the right of the most significant digit. The second factor is the power of ten, and only whole integers are used as powers. For example, the speed of light in a vacuum to six significant figures would be written as 2.99796×10^8 m/sec.

Further reference to error analysis and its presentation in the technical report is available in some of the works listed for your convenience under SUGGESTED REFERENCES.

SUGGESTED REFERENCES

American Society of Mechanical Engineers, *American Standard Engineering and Scientific Graphs for Publication,* 1947.

ASTM, *ASTM Metric Practice Guide,* E 380-72, 1972.

Beers, Y., *Introduction to the Theory of Error,* Addison-Wesley Publishing Co., Reading, Mass., 1953.

Blicq, R.S., *Technically Write!* Prentice-Hall, Inc., Englewood Cliffs, N.J., 1972.

Bragg, G.M., *Principles of Experimentation and Measurement,* Prentice-Hall, Inc., Englewood Cliffs, N.J., 1974.

Schenck, H., Jr., *Theories of Engineering Experimentation,* McGraw-Hill Book Co., New York, 1961.

Wirkus, T.E., and Erickson, H.P., *Communications and the Technical Man,* Prentice-Hall, Inc., Englewood Cliffs, N.J., 1972.